Contents

What is rain?

Rain is water that falls from clouds. Clouds form in the sky where it is colder than the ground. Each cloud is made up of billions of water droplets.

Rain is falling from this cloud into the sea.

rain

4

What is Weather?
RAIN

Andy Owen and
Miranda Ashwell

Heinemann
LIBRARY

First published in Great Britain by Heinemann Library,
Halley Court, Jordan Hill, Oxford OX2 8EJ,
a division of Reed Educational and Professional Publishing Ltd.
Heinemann is a registered trademark of Reed Educational & Professional Publishing Limited.

OXFORD MELBOURNE AUCKLAND
JOHANNESBURG BLANTYRE GABORONE
IBADAN PORTSMOUTH NH (USA) CHICAGO

Designed by David Oakley and Celia Floyd
Originated by Dot Gradations, UK
Printed and bound in Hong Kong/China.

04 03 02 01 00
10 9 8 7 6 5 4 3 2 1

ISBN 0 431 03782 5
This book is also available in hardback (ISBN 0 431 03777 9).

British Library Cataloguing in Publication Data

Owen, Andy
Rain. - (What is weather?) (Take-off!)
1. Rain and rainfall - Juvenile literature
I. Title II. Ashwell, Miranda
551.5'77

Acknowledgements
The Publishers would like to thank the following for permission to reproduce photographs:
Bruce Coleman Limited: E Bjurstrom p28, M Boulton p26, J Cancalosi p27, A Compost p18, S Krasemann p11, F Prenzel p29, H Reinhard p25, K Rushby p21; FLPA: C Mattison p19; Robert Harding Picture Library: pp4, 6, 7, J Francillon p30, J Miller p9; Oxford Scientific Films: S Olwe p15, K Wothe p17; Andy Owen: p8; Panos Pictures: J-L Dugasi p22; Planet Earth Pictures: R Salm p16; Tony Stone Images: P Cutler p5; Still Pictures: M Edwards p20, P Gleizes p10, M Gunther p12, R Pfortner p24, A Watson p13; Telegraph Colour Library: C Mellor p14, A Mo p23.

Cover: The Stock Market/Bruce Peebles
Our thanks to Sue Graves and Stephanie Byars for their advice and expertise in the preparation of this book.

Every effort has been made to contact copyright holders of any material reproduced in this book. Any omissions will be rectified in subsequent printings if notice is given to the Publisher.

For more information about Heinemann Library books, or to order, please telephone +44 (0)1865 888066, or send a fax to +44 (0)1865 314091. You can visit our website at www.heinemann.co.uk

Any words appearing in the text in bold, **like this**, are explained in the Glossary.

When lots of rain falls in a short time, we say that it is raining heavily. Heavy rain does not soak into the ground very quickly so sometimes puddles form on the ground.

Did you ever splash in puddles, like these, when you were small?

Why does it rain?

The air is full of tiny droplets of water but they are too small to see. Sometimes there are so many droplets that the air feels damp. We might even see some **mist**. Mist forms when the ground is cold. You will often see mist early in the morning.

Mists often form on cold, autumn mornings.

Each of these small clouds has many water droplets in it.

When the air cools, the water droplets join together. These larger water droplets make clouds in the sky which we can see. When the water droplets get too heavy to stay up in the air, they fall out of the sky as rain.

Where does it rain?

Rain often happens in areas where there are mountains. The wind carries the clouds over the mountains where the air is cold. The cold air turns the water droplets in the clouds into falling rain.

The wind has carried this cloud high up over the mountain.

Rainforests are places that get heavy rainfall.

It also rains when damp air cools. In **rainforests** it rains nearly every day. The rainfall is heavy. The warm, damp air rises. As it rises it cools, making clouds and heavy rain.

Fog and hail

When clouds lie near to the ground **fog** can form. Fog forms when air near the ground cools. Driving in fog on a motorway can be dangerous because it is hard to see. In fog you can see less than 1 km in front of you.

Driving in thick fog on a motorway can be dangerous.

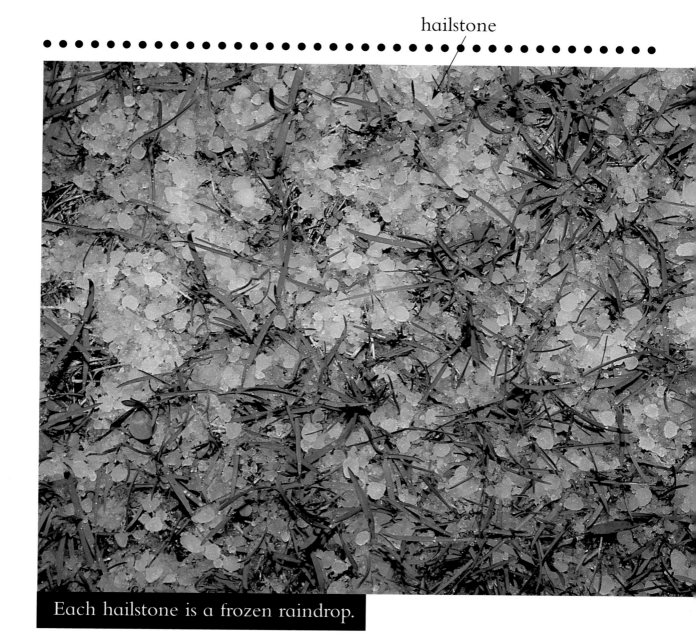

hailstone

Each hailstone is a frozen raindrop.

Hail falls when raindrops in the clouds get very cold. The raindrops freeze into hard pieces of ice and fall to the ground as hail.

11

Wet or dry

Rainforests have more rain than anywhere else in the world. It rains nearly every day in these hot, wet forests.

Forests, like these, are called rainforests because they have more than 2000 mm of rain every year!

Deserts are always very dry.

It may not rain for years in a **desert**. Deserts can be hot or cold, but they are always very dry places.

The highest temperature recorded in the Sahara Desert was 84°C (151°F).

Rain patterns

In some places, many months go by without any rain. Then, at the same time every year, rain falls. This is called the **monsoon season**.

The streets in this city in India can be hot and dusty for many months.

The monsoon rains flood the city streets.

When the monsoon season starts, the rain falls very heavily. It lasts for many months. People use the rain-water to help grow their food.

In India the monsoon season starts in June and can last for four months.

Life in dry places

All animals need water to live. This camel lives in the **desert** where there is very little rain. It can live for a long time without food or water because it can store fat in its hump. The camel's body can change the fat into water when it needs to.

A camel must drink sometimes!

Cactus plants can survive in hot, dry deserts.

Plants need water to grow or they will **wilt** and die. Cactus plants are used to living in deserts. Cactus plants store water inside their thick **stems**.

People crossing the desert can cut into a cactus to get water.

Life in wet places

Plants in the **rainforest** get plenty of rain. The rain and sunshine help the plants to grow very large. They also grow very quickly.

Plenty of rain helps these plants to grow.

The leaves of rainforest plants get very wet. Rain flows along the shiny leaves. Frogs live in the tiny pools of water held by the leaves.

This frog lives in a tiny pool of water on a leaf.

Drinking water

Everyone needs drinking water. Drinking water comes from rain-water. Our homes have taps so that we can get water whenever we want. But in some hot countries, water is **scarce**.

These people are collecting water from a village pump.

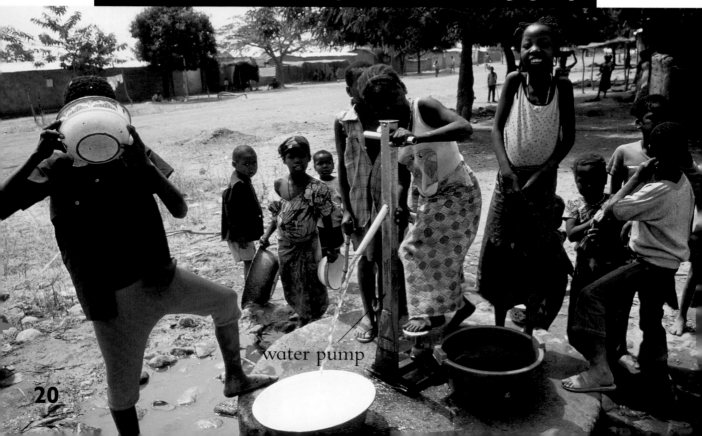

water pump

Drawing water from the well.

When it rains, rain-water soaks into the ground. A hole, called a well, has been dug to reach the water deep in the ground. People draw water from the well for themselves and also to give to their animals.

Collecting rain-water

Our rain-water is held in **reservoirs**. When it has not rained for many months, there is not much water left in the reservoirs. People have used most of the water. When there is not much water left in the reservoirs, people are banned from wasting water.

This reservoir has no more water left in it.

Clean water is piped into our homes from the water works.

After weeks of rain the reservoir is full of water. The water is piped to places called water works where it is cleaned. Then the water is piped into our homes and **factories**.

Acid rain

Smoke from the chimneys of homes and **factories** makes the air dirty. Some of the dirt gets into water drops and falls with the rain. We call this **acid rain**.

Smoke like this, from factory chimneys, makes the air dirty.

bare branches

stripped bark

Acid rain has killed this forest.

Acid rain harms many living things. These trees are dying because acid rain has fallen here for many years.

Acid rain has killed many forests in Europe and parts of North America. It also damages buildings.

Too much rain

Rivers flow with rain-water. After heavy rain, water can spill out over the river's banks and onto the land. We call this a **flood**.

This land has flooded after heavy rainfall.

Flood-water has poured into this house.

Floods can be very dangerous. The flood-water can flow quickly. It can flow into houses and people can be trapped. Fast-flowing water can sweep away anything in its path. Animals, too, can be at risk when there is a flood.

Too little rain

When there is no rain for some time, the soil dries out and becomes cracked and hard. Most plants cannot live for long without water. They soon **wilt** and die.

Plants quickly wilt and die without rain.

cracks in the dried-up soil

A **drought** is when it does not rain for a long time. Droughts usually happen in hot, dry countries but they can happen anywhere. Even Britain has droughts!

Animals may die if they cannot find drinking water in a drought.

It's amazing!

Sometimes we see lightning when it rains. Lightning is a huge spark of electricity which lights up the sky. Thunder rumbles as the air heats up quickly.

There are more than 16 million thunderstorms in the world every year.

Glossary

acid rain   rain that has become mixed with smoke from factories and cars. It can damage plants and buildings

desert   a place where there is very little rain

drought   a long time without rain

factory   a place where people make things

flood   when the land is covered by water

fog   when the air is so full of drops of water that it is difficult to see very far

hail   rain that falls as drops of ice

mist   when it is damp and cold, we can sometimes see the drops of water in the air

monsoon season   heavy rain which comes at the same time every year

rainforest   thick forest which grows in hot, rainy places

reservoir   a place where water is stored

scarce   not very much of something

stem   the part of a plant from which the leaves and flowers grow

wilt   when plants have no water their leaves become limp and they die

Index